American Breeds of Beef Cattle
With Remarks on Beef Cattle Pedigrees

by US Dept. of Agriculture

with an introduction by Jackson Chambers

This work contains material that was originally published in 1902.

This publication is within the Public Domain.

This edition is reprinted for educational purposes and in accordance with all applicable Federal Laws.

Introduction Copyright 2017 by Jackson Chambers

Self Reliance Books

Get more historic titles on animal and stock breeding, gardening and old fashioned skills by visiting us at:

http://selfreliancebooks.blogspot.com/

Introduction

I am pleased to present another title in the "Cattle" series.

The work is in the Public Domain and is re-printed here in accordance with Federal Laws.

As with all reprinted books of this age that are intended to perfectly reproduce the original edition, considerable pains and effort had to be undertaken to correct fading and sometimes outright damage to existing proofs of this title. At times, this task is quite monumental, requiring an almost total "rebuilding" of some pages from digital proofs of multiple copies. Despite this, imperfections still sometimes exist in the final proof and may detract from the visual appearance of the text.

I hope you enjoy reading this book as much as I enjoyed making it available to readers again.

Jackson Chambers

ORGANIZATION OF THE BUREAU OF ANIMAL INDUSTRY.

Chief: D. E. Salmon, D. V. M.
Assistant Chief: A. D. Melvin, D. V. S.
Chief Clerk: S. R. Burch.
Dairy Division: Henry E. Alvord, C. E., chief; R. A. Pearson, M. S., assistant chief.
Inspection Division: A. M. Farrington, B. S., B. V. S., chief; E. J. Jones, assistant chief.
Miscellaneous Division: Richard W. Hickman, Ph. G., V. M. D., chief.
Editorial Clerk: George Fayette Thompson.
Artist: W. S. D. Haines.
Expert in Animal Husbandry: George M. Rommel, B. S. A.
Librarian: Beatrice C. Oberly.

LABORATORIES.

Biochemic Division: E. A. de Schweinitz, Ph. D., M. D., chief; Marion Dorset, M. D., assistant chief.
Expert in Dairy Chemistry: George E. Patrick, M. S.
Pathological Division: Ch. Wardell Stiles, M. S., Ph. D., A. M., chief.
Zoological Division: John R. Mohler, V. M. D., zoologist; Albert Hassall, M. R. C. V. S., acting assistant zoologist.

EXPERIMENT STATION.

Superintendent: E. C. Schroeder, M. D. V.; expert assistant, W. E. Cotton.

LETTER OF TRANSMITTAL.

U. S. DEPARTMENT OF AGRICULTURE,
BUREAU OF ANIMAL INDUSTRY,
Washington, D. C., October 1, 1901.

SIR: I have the honor to submit herewith the manuscript of an article on "American breeds of beef cattle, with remarks on pedigrees," by George M. Rommel, expert in animal husbandry of this Bureau, and recommend that it be published as Bulletin No. 34 in the Bureau series.

Respectfully,
D. E. SALMON,
Chief of Bureau.

Hon. JAMES WILSON,
Secretary of Agriculture.

CONTENTS.

	Page.
Introduction	7
The awakening	9
American improvement	10
Beef type	11
What is a breed?	13
The breeds of beef cattle	13
Shorthorns	14
Herefords	15
Aberdeen-Angus	17
Galloways	18
Devons	19
Sussex	20
Red Polls	20
Polled Durhams	21
Brown Swiss	21
Polled Herefords	21
Pedigrees	22
The condensed form	22
The tabulated form	24
What constitutes a good pedigree	24
Eligibility	24
The breeder	26
Merit	26
Families	27
The value of a good family	27
Registration	28
Certificates of registry	28
Miscellaneous documents	29
Certified pedigree	29
Transfer	29
Breeding certificates	29
Fees	29
The herdbook	30
Shorthorn	30
Hereford	31
Aberdeen-Angus	32
Galloway	32
Devon	32
Red Poll	32
Polled Durham	33
Breeders' organizations	34

ILLUSTRATIONS.

		Page.
Plate I.	The Blackwell Ox	8
II.	Advance	8
III.	Hereford Bull Dale	8
IV.	Shorthorn Heifer Ruberta	8
V.	Shorthorn Bull Viscount of Anoka	12
VI.	Hereford Bull Sir Bredwell	12
VII.	Rear view of Viscount of Anoka	12
VIII.	Rear view of Hereford Bull Sir Bredwell	12
IX.	Shorthorn Cow Mary Abbotsburn 7th	16
X.	Hereford Cow Betty 2d	16
XI.	Aberdeen-Angus Bull Orin of Longbranch	16
XII.	Galloway Bull Speculator of Dundee	16
XIII.	Front and rear view of Orin of Longbranch	20
XIV.	Galloway Heifer Gentle Annie A	24
XV.	Galloway Heifer Lutie Lake	24
XVI.	Devon Bull Tulip's Royal 1st	24
XVII.	Devon Heifer Tulip of Woodland 23d	24
XVIII.	Red Polled Bull Demon	32
XIX.	Polled Heifer Lida Falstaff 3d	32
XX.	Polled Durham Bull Golden Gauntlet	32
XXI.	Polled Durham Cow Goodness 15th	32
XXII.	Polled Hereford Bull	32
XXIII.	Polled Hereford Cow	32

AMERICAN BREEDS OF BEEF CATTLE.

INTRODUCTION.

Prior to the discovery of America there were no cattle in the Western Hemisphere. On one of his voyages to America Columbus is said to have brought a number of domestic animals with him. Succeeding Spanish explorers followed his lead, and each military expedition that had cavalry as a part of its equipment added to the supply of horses. As the conquerors mingled more closely with the natives and settled down to peaceful pursuits or wandered into the interior, it was but natural that, with the multiplication of domestic animals, many escaped to establish themselves in favorable environment. For these there were grass and water in abundance. There was the opportunity for sowing the seed that produced the native cattle of the West Indies and Mexico, the long-horned herds of Texas, and the wild horses of the Plains.

The start in the South was followed about the middle of the sixteenth century by the Portuguese, who took cattle to Newfoundland and Nova Scotia, and by the introduction of cattle into Acadia and inland New France by the French. These were carried farther into the interior, and it is said that the French missionaries in Illinois about 1750 possessed considerable numbers of cattle, horses, and swine. Cattle were introduced into Virginia shortly after the Jamestown settlement and multiplied rapidly. Legislation which made their killing a crime punishable with death contributed to this result. A bull and three heifers were brought to Plymouth by the ship *Charity* in 1624. New Hampshire received cattle from Denmark, New Netherlands from Holland, and Delaware from Sweden, about the same time. Carolina received her first shipment from England in 1670, while Georgia was the last of the colonies to figure as a market for the English export trade in breeding cattle.

It is a reasonable inference that these cattle represented the best stock in the countries from which the settlers came, but as these first importations were mostly made over one hundred years before the English improvement in cattle breeding set in, they were, of course, of an unimproved type. It is a note of interest that the ship *Charity* sailed from a Devonshire port, and that many other ships to New England in the first few years following sailed from Devon, which, supplemented by the reputation of the "red cattle of New England" that has come down to us, makes it quite possible that these were unimproved Devons. The New Netherlands importations were undoubtedly of a milking breed, but all were in time so hopelessly intercrossed that their identity was lost, and, as a result, our forefathers had the "native cattle" of the Eastern United States.

Such was the foundation. What the native stock was like we can best imagine from the stories of men now old, and from the "scrub" stock that is even yet the eyesore of many American pastures. Blood of Spanish, Swedish, French, Dutch, and English, with, maybe, a dash of buffalo as they wandered westward, gave this stock a cosmopolitan character that was representative, perhaps, but hardly profitable. Lack of care by farmers, with no Bakewell to point the way to improvement, brought about a type of animal that a century has not been able to absorb.

In England and Scotland the situation was slightly better. The age of the country and the scarcity of land made possible and necessary a more thorough system of agricultural operations, and had early brought out the predominant influence of live stock in a well-regulated system of husbandry. However, compared with what we see to-day, conditions were embryonic and crude. Huge, rough, ungainly beasts were the animals whose descendants were to be the winners at Smithfield and Chicago. A fair idea of the cattle that were the typical native stock of Great Britain up to the middle of the eighteenth century may be formed by a study of the noted "Blackwell ox." He was over five years old when killed at Darlington, December 17, 1779.

Weight and dimensions of the Blackwell ox.[a]

WEIGHT.

	St.	lb.	[Pounds.]
Fore quarters	76	7	[1,071]
Hind quarters	76	3	[1,067]
Tallow	11	..	[154]
Total	163	10	[2,292]

DIMENSIONS.

	Ft.	in.
Height at crop	6	...
Height at shoulders	5	9¼
Height at loins	5	8
Height from breast to ground	2	1
Length from horn to rump	9	3½
Breadth over shoulders	2	10½
Breadth across hips	2	10½
Girth before shoulders	9	7½
Girth behind shoulders	10	6
Girth at loins	9	6½

[a] Transactions of the Highland and Agricultural Society of Scotland, 1896.

These figures, with the illustration (Pl. I), show an animal of good grazing qualities, strong constitution but light in flesh-carrying capacity, much coarseness, and unquestionably a bad feeder. This was the old Teeswater type, from which the Shorthorn breed was descended, and represents a period about midway between the beginning of Bakewell's work and that of the Collings.

Things were forming, but not formed. There was no fixed type and no method yet in general use by which a type could be readily fixed.

THE AWAKENING.

It was not until near the end of the eighteenth century that the historical period of pure-bred beef cattle began. The years from 1760 to 1837 were the formative stage of Anglo-Saxon cattle breeding; it was about the year 1760 that Bakewell's operations began. The year 1837 marks the beginning of Amos Cruickshank's tenantry at Sittyton. Around the name of Robert Bakewell those of all great improvers of live stock group themselves, and from the lessons that he taught by example, if not by precept, every breeder learns the fundamentals of his art. Previous to him we find a class of cattle of no uniformity, of little value as high-class meat producers, late maturing, and without quality. After his time, we see an era of wonderful growth and improvement. A man of marked attainments, striking out for himself, he achieved results, by close study of anatomical structure and heredity, that changed breeding methods the world over. He was the first man to practice systematic inbreeding, the stock used being Leicester sheep and Longhorn cattle. He was a constant student—a great man—learning new facts by means of experiment and comparison, always keeping in view the most economical utilization of every force and product of the farm. Though he kept his methods to himself to a great extent, the great fact of them all—that the surest way to improve stock is by the use of inbreeding in the hands of a master—serves to perpetuate his name. His work was the start, and before the end of the eighteenth century the Collings and Thomas Booth with the Shorthorns, and Tompkins with the Herefords, had crystallized the bloods of the popular strains in York and Durham and Herefordshire; while a few years later, at Keillor, Hugh Watson started the "Doddies" on their triumphant career; Quartley was making his a family name among Devon ranks; the sturdy Scotch farmers of Galloway were working slowly but surely to bring their cattle to the high standard set in the South; and the first year of the nineteenth century witnessed the advent of Thomas Bates on the Shorthorn field of action.

While Bakewell showed the way to improvement, Cruickshank emphasized the necessity of utility in connection with pedigree and, saving a great breed from almost certain ruin, formulated that idea as a principle that is coming to be more and more regarded as the cornerstone of successful breeding operations.

Out of the dark ages of ignorance and of the scrub, by leaps and bounds, using what material he had at hand and molding it to his will, the English farmer developed the modern breeds of cattle; producing tender meat where tough and leathery fiber had been before, paying the rent with his cattle and sheep, and, in time, contributing very largely to the growth of agriculture in the New World. Americans are justly proud of their cattle. Their breeders have reached a point up to if not beyond that of their contemporaries across the water; but no supremacy of excellence, no victory in show ring or market, can efface the memory of the debt America owes to those sturdy yeomen whose names adorn the herd records of England and Scotland.

AMERICAN IMPROVEMENT.

Improvement in America began almost simultaneously with that in England. No sooner had the Revolutionary War closed than importations began, continuing at frequent intervals until the outbreak of the war of 1812. These cattle were mainly of Shorthorn breeding and were distributed along the coast from New York to Virginia, some finding their way into Kentucky and Ohio.

The year 1817 will always be a memorable one in American cattle history. Following the Cox importation of Shorthorns, immediately after the close of the war of 1812, this year marks the beginning of the Devons and Herefords, another importation of Shorthorns, and the stillborn advent of the Longhorns. The Shorthorn importation of this year, and the previous Cox importation, are notable for another reason, as both mark the initial point of Shorthorn pedigree folly in the United States.

Growth was slow for ten years, when renewed activity was evidenced, especially in Shorthorns. Companies were formed and the improvement in cattle was marked. In point of numbers the Shorthorn breed rapidly assumed a foremost position, and until about the year 1880 was the only beef breed of prominence

The expansion of the cattle business was rapid. Up to the opening of the transcontinental railways it was mainly carried on in the sections east of the Missouri River, but with the entrance of these arteries of trade into the plains and the discovery of the great opportunities of this country for grazing, a growth set in that was too rapid to be normal. In the early eighties pure bred cattle by the thousands were brought from England to supplement the American herds in breeding bulls for the range, and the nearest that the Hereford and Angus breeds ever came to having a boom in this country was at this time. After the collapse, which was bound to follow, the cattle business is now on what is thought to be a substantial and healthy foundation. Quality is being bred into the range herds by the extending use of pure-bred sires, and this, with better methods, is bringing the range steer to a high plane of excellence. Both on the range and on the small farm, improvement has gone hand in hand with increase in numbers.

In taking up an examination of the breeds of beef cattle in detail, keeping their characteristics in mind so that a just comparison may be reached, we find that in all important points, considering them simply as finished beef animals, ready for the block, there is little difference between individuals of the various breeds. There is a tendency for all to approach a common standard. All have been developed with a common object in view—the production of beef. This result will be accomplished by practically the same methods the world over, varying only on account of the facilities at hand and market demands, and this breeding of cattle solely for the production of beef has resulted in the formation of the "beef type." Carrying the idea still further, we may say that breeding the same species of animals over a long continued time for a single purpose results in the formation of a "type" whose characteristics are fairly well fixed, the word "type" being arbitrarily used for lack of a more convenient one. Thus, in cattle, in addition to the beef type, we shall find that breeding for the special development of the milking function has resulted in a fairly well-defined dairy type; in horses, breeding for the different purposes of speed and draft has resulted in two very well-defined types; while the same result has, to a large extent, been brought about among the breeds of sheep by breeding specially for mutton or for wool. In addition to the influence of the breeder in the development of a type, the influence of soil and climate has been great.

This being true, the longer the breeds of one type are bred to meet the demands of a common market the nearer will they approach the common standard, and this is especially so in the case of animals which are a source of food supply where fashion and taste can not prevail in the face of dollars and cents. This general type is what the farmer looks for in selecting his feeding cattle, and, with very many men, this is all they consider, the matter of breeding being looked upon as of minor importance. The knowledge of this framework is the first essential and, unless thoroughly learned, one can not lay claim to a knowledge of the breeds.

BEEF TYPE.

The first point observed in an ideal animal of beef type is his form. This will approximate the rectangular. It will show a body that is compact, symmetrical, broad, deep, and close to the ground. Legs are only of use to carry the animal around. He is "straight in his lines;" that is, the lines from the top of the shoulder to the tail head, and from the brisket back to the purse, are as nearly parallel as possible, as are also those from the center of the shoulders to the center of the thighs, no deviation from the horizontal being allowed the top line. This will give the form a rectangular appearance.

The head shows a "good feeder." Observation and experience show a good feeding head to be broad and well filled between the eyes, with a good roomy brain box, tapering nicely, and short from the eyes to the muzzle, which should be wide and clean cut, with large, well-open nostrils. A large mouth usually is the first indication of good digestive capacity, and large, open nostrils go with good lung power and a strong constitution. The whole head is clean cut, with no superfluous flesh on the jaws. The horns, if present, are of medium size and not coarse; ears of medium size, gracefully and actively carried. The eyes are large, full, bright, clear, and placid. The neck of most animals of extreme beef type is reduced to the shortest degree possible with usefulness. It is moderately full, with clean-cut throat, large, well-defined windpipe, and little or no dewlap. The neck joins the shoulder in full, even lines, swelling into the shoulders, as it were.

Shoulders are moderately sloping, well covered with flesh, smooth, deep, and wide; tops of shoulders compactly covered with smooth flesh. The chest wide and deep with the crops full and heart girth full. The brisket will be wide and projecting forward, carrying a moderate amount of flesh. The ribs are well sprung, long, and close together, giving the animal plenty of room for the work of the organs of the chest and abdomen. The space between the last ribs and the hips is short, giving the body a compact appearance.

Thus far we have concerned ourselves principally with the parts that have to do with the organic nature of the body. Along the back, from the shoulders to the hips, we find the most valuable meat in the animal's carcass. Here, then, we have the greatest width and depth of flesh possible. The spring of the ribs mentioned, the length and breadth of back and loin, and, above all, the depth of flesh on ribs, back, and loin, are absolute essentials of the beef type.

The hips are of medium breadth, well covered and smooth, the rump long, wide, and well filled in from hips to tail head. The thighs are wide, deep, and full, and the same description applies to the twist. The flesh of the hind quarters is carried well down to the hocks. Legs are short, straight, and strong, with fine, clean bone, and set well outside the body.

Above all, every part of the body of an ideal beef animal shows "quality." It is this that tells the farmer whether a steer with a good form will prove a profitable feeder; it tells the butcher whether the animal will "kill" well. This is a point that can not be overlooked and is difficult to describe. Lack of it shows in a coarse, fleshy head, in a thick, meaty throat, and a rough, uneven shoulder. Coarse, heavy bones and a loose-jointed appearance generally will show the undesirable feeder; rough flesh, "ties," and "patches," the undesirable killer. The animal handles well. The flesh is mellow and firm, showing a proper mixture of fat and lean. The skin is loose, but not superfluous, mellow, and moderately thick, covered with a plentiful growth of hair. Such an animal usually weighs 1,500 pounds at twenty-four to thirty months.

WHAT IS A BREED?

A type may be defined as the result achieved by breeding domestic animals for a fixed and definite purpose. Definite breeds of the same type have been evolved by the development of animals for the same purposes, but with slight differences that give slightly different results. These differences—the influences that go to make up a breed—will be (1) environment, (2) the characteristics of the original stock, and (3) personal preferences of the breeders. The outstanding work of one man or the standard set by many will gradually fix certain characteristics until they become practically constant, and a breed is the result. That breeds as we have them have been developed solely by man's efforts is shown by the fact that, if neglected, all breeds of one species in the same locality tend to revert to a common type—the original species or variety from which all have been developed.

THE BREEDS OF BEEF CATTLE.

On leaving the general beef type, which is common to all beef cattle, and taking up a consideration of the breeds in detail, the first marked difference that presents itself is the requirement that breeding cattle possess what is generally known as "character." In the preceding description we had before us the fattened steer. Character must be present in fat stock, but the word is not used with the same application as with breeding stock. A steer's business is to turn himself into marketable beef of the best quality in the shortest possible time. If the farmer knows a steer will do this he is satisfied. Character shows in the lower animals somewhat as it does in man, and the head is the mirror that reflects it. In a bull that which impresses one with a sense of dignity, individuality, and power, with an unmistakable masculinity, is his character; on the other hand, we note character in a cow by her individuality, dignity, and femininity—a sweet, big-hearted motherliness—that, no doubt, will cause any calf to run to her. We can not tell what that power is anymore than in human beings. It is there; we feel its presence; still more, we note its absence. Experience must teach it.

The bull must be masculine; he must have a strong crest and a bull's head. The cow must be feminine, with no trace of masculinity. In addition, breeding stock must be prepotent, able to transmit their characteristics accurately and uniformly. This can hardly be estimated by any external indication. A strongly prepotent animal will usually have a great deal of character, but the only sure indication is the animal's work as a breeder. Prepotency is thought to be hereditary as well as feeding qualities, and close study of a pedigree is a fairly satisfactory way to determine how a youngster will develop.

We may summarize, then, the qualities that we find common or necessary to all breeds of beef cattle as follows:

(1) Character and sexual characteristics—prepotency.

(2) Beef type—"carrying the greatest amount of flesh in the smallest superficies."

(3) Quality.

(4) Early maturity—ability to produce the greatest weight of prime flesh in the shortest time.

Observe that all but the first are brought about by market demands.

SHORTHORNS.

This is one of the heaviest of the beef breeds. Mature bulls in show shape often attain a weight of 2,700 pounds and cows 2,000 pounds. However, these extreme weights, without the highest quality, are objectionable. The color is more variable than that of any other breed; it may be red or white, or a mixture of these colors, the colors popularly used to describe the breed thus being red, white, and roan. Roan "is, indeed, the one distinctive Shorthorn color never produced except by the presence of the blood of this breed."[a] Taking up a detailed examination, note the width and depth of form, its great scale and substance, and the general impression of style. We see an animal possessed of quality, a clean-cut appearance straight through, the body set upon legs of medium length with a clean bone of moderate substance. In the head, observe the width between the eyes—the fullness of the brain box and the expression of great character. A typical Shorthorn head will afford a man weeks of study in bovine character. The indications of a good feeder are also present. In passing, note the short horn, curving gracefully forward and occasionally drooping, waxy and white in color with black tips. The neck shows strength and sexual power, and is joined to a rather upright shoulder by a smooth and full shoulder vein. Passing to the body, note the heavy flesh and the spread of back and loin. In the hind quarters especially Shorthorn characteristics are present. Indeed, the breed has the reputation of carrying the best hind quarter of any. Note the width of hips, the length and width of rump. Further on, the

---

[a] Sanders' Shorthorn Cattle, p. 14.

great width, depth, and fullness of thigh and twist and the way that the flesh is carried right down to the hock are prominent features. Especially characteristic is the line of the back of the thigh. This is nearly straight from the tail down, making the plane of the thighs nearly level. Legs are of moderate length, with a bone of medium fineness and plenty of strength.

Three groups of Shorthorn cattle have been evolved during the last century. During the first fifty years, the Booth and Bates families were developed and the opposition between them was as great as between different breeds. The Booth cattle were famous for their fleshing qualities. Bates cattle were famous for both milk and beef production. They grazed well and possessed high quality and much style. Later the development of the Scotch sorts, under the guidance of the Cruickshanks, brought forth animals of a blocky, short-legged type, with much scale and substance, excellent fattening powers, good constitution, quality, and early maturity. The tendency during the past thirty years has been to combine the blood of the old strains.

As individuals, the weak points of Shorthorns are a tendency to be rather long of leg, flat in the ribs, and light in heart girth, and an undue prominence of the hook bones, with a consequent lack of fullness of the rump. On the other hand, the great flesh-carrying power, development of hind quarter, quality, and prepotency are manifest.

As a breed, its popularity and wide dissemination have given rise to variation in types that has resulted in many a show-ring dispute; but we have, to offset this, the wonderful adaptability of the breed, its seeming universal usefulness, and its value in crossing with almost every other. Shorthorn blood was the first to be used on the native cattle of the Plains, and exercises a very great influence on the range cattle of the present time.

An estimate of 150,000 as the number of living registered Shorthorns in the United States is approximately correct. Of these, it is stated that 5 per cent are on the range and 95 per cent in the hands of the "small farmer," the strong feature of the breed being its adaptability to the requirements of diversified farming.

HEREFORDS.

In weight, Herefords are about equal with the Shorthorns. The extreme weights to which Shorthorn bulls frequently attain are rare, but, generally speaking, there is practically no difference between the two breeds.

Hereford color is easily recognized. That most favored is a rich medium red with white markings. Note the extent of white; head and face, top of neck, dewlap, brisket, belly, front feet, and hind legs below hocks, and brush of tail are usually white. This arrangement is not absolute. A streak may be present on the middle of the back, and the white may extend over a very much larger area, even to the extent of making a spotted effect. Ears are usually red or spotted, rarely solid color; red spots often are present on the head, especially around the eyes. The red may vary from a light yellow red to dark, almost black in some animals. Time was when the face was gray, or mottled, in some animals and spotted in others. Even an occasional white animal was met with. The advent of a spotted calf in a herd is therefore no indication of impurity.

In the Hereford the width and lowness of the form is extreme. Close to the ground, broad, blocky, deep, nicely rounded, and stylish, the Hereford is typical. The head is a splendid feeder's pattern—broad, short, and full of character, with a capacious mouth and large nostrils, showing good digestion and strong constitution. The muzzle is light in color, without spots. The horn is white, somewhat coarser and considerably longer than that of the Shorthorn, has longer curves, sometimes being nearly straight, and usually has a drooping tendency, especially in the best-bred cattle. In cows and steers it is often elevated, but this is rare in bulls.

Observe how closely the head is set to the body—a noteworthy point of economy. The neck is cheap meat; therefore, eliminate the neck if you can. The development of the fore quarters, or "fore hand," the width of chest and heart girth have been objects of special care by Hereford breeders. Accompanying these will be the strong constitution that has given the Hereford his hold on the range. Back, loin, and ribs carry a tremendous amount of flesh of fine quality. The Hereford hind quarter has been somewhat differently developed from that of the Shorthorn, and has been remarkably improved during the last twenty-five years. The square-cut, packed-in-a-box appearance that the Shorthorn presents is missing here. The hips are not quite so wide nor prominent as in the Shorthorn, generally a little smoother, the rump wide and well filled, and, instead of a straight quarter behind, we see a slightly bulging one, more so than in the Shorthorn, but less than in the Angus.

Individually, the Hereford tends sometimes to coarseness, with a light hind quarter, but he has in his favor a very compact body, a deep fore hand, large well-rounded heart girth, and great depth and spread of flesh on back, rib, and loin. It is sometimes claimed that the Hereford lacks scale and size. If this was true at one time, figures have conclusively proved that it is not now.

As a breed, their long specialization for beef production has operated to give them a less general distribution than the Shorthorns; but the strength of the breed lies in this very fact. Its value for grazing purposes and for prime beef production commands attention. The close uniformity of type is also noteworthy.

Though among the first breeds to be introduced, Herefords were little known in the United States until the opening of the range country. The impetus which this gave the cattle trade brought them into prominence, and, beginning about the year 1880, a rapid growth has put the white faces second in point of numbers in this country and almost supreme on the range.

It is estimated that the number of registered Herefords in the United States at present is about 70,000. Of these, about 14 per cent are in the range States.

ABERDEEN-ANGUS.

Examining an Angus bull for the first time, we shall notice first his black color. He is all black, the only white allowed being a little on the underline behind the navel. Some white on the udder is not objectionable, as it is thought to be usually present with the best milkers, but white is not wanted on the cod. Occasionally red calves are dropped, showing a tendency to revert to the animals of the eighteenth century, when mixed colors were comparatively common. Having observed the color, we note an entire absence of horns. Not even scurs are allowed.

Going more into detail, note next the form. This presents a considerable variation from the ones previously discussed, and is very typical of the breed. While the requirements of a first-class beef animal demand a rounded form, here we find this carried to an extreme, and the "barrel shape" is a characteristic that the Angus claims peculiar to itself. Viewed from any direction, this marked rotundity is prominent; and one notices, too, how low-set the animal is, his great style, quality, compactness, and symmetry. In the feeder we find all the indications of a good beef-making machine and in the finished animal every requirement of a market topper. We see short legs, and neat, fine bone, and most particularly the wonderful smoothness which even the most extreme forcing is hardly able to mar.

The head is very characteristic. Short, wide, clean cut, with a muzzle whose capacious mouth and large nostrils denote excellent feeding qualities and strong powers of constitution; surmounted with a tufted poll that is sharp and higher in the female than in the male, and ornamented with eyes of rare beauty, and large, hairy ears, elegantly carried; the whole set to the body with a neck almost as short as that of a Hereford. The Angus head is an index of the excellence that we are to find behind it.

Note the tremendous width of chest, with legs set well outside the body, the spring of rib, and deep heavy flesh. Observe the compactness, how closely the ribs are joined to the hind quarters. In the hind quarters we fail to find the hook bones. They are there, but so well concealed by smooth flesh that frequently the most careful handling fails to locate them. Here we find still other Angus characteristics. The tail is set a little farther forward than in the Shorthorn. The buttocks are more rounded, but the quarter carries a large amount of flesh well down to the hocks.

Individuals of this breed do not attain the great weights of the Shorthorns, mature bulls rarely weighing over 2,200 pounds, and cows averaging perhaps 1,600 pounds in show condition. But early maturity enables them to attain marketable weights in an extremely short time. Angus bulls are strongly prepotent, getting calves of great uniformity, from 75 to 90 per cent of which from horned cows are polled.

The milking qualities of Angus cows have been considerably neglected, and all the powers of the breed have been directed to the production of prime beef. How well this has been accomplished the markets and fat-stock shows both in England and America will witness. They stand heavy feeding admirably, and the bulls are in use all over the country in grade herds.

The principal defects of cattle of this breed will be found to be a tendency to coarse shoulders and narrow rumps. Special points of merit are their quality, early maturity, and excellence for the block. As a breed, they are prepotent and uniform in type.

The first importation of Angus cattle into the United States was that of three bulls brought over in 1873 for use on the native cows of Kansas and Indian Territory. The offspring of these bulls attracted much attention, and subsequent importations soon made the breed well known. They shared with the Herefords the rush of the early eighties, it being estimated that over 2,000 head were imported from 1879 to 1882. There are at present about 40,000 registered animals living in North America, about one-fifth of which are on the range, the other four-fifths being in the hands of the small farmer and breeder. Their footing on the range has always been secure, but the popularity of the breed for the production of "baby beef" keeps most of them east of the Missouri.

The outlook is bright for Angus cattle in America. Under date of July 15, 1901, Secretary Thos. McFarlane, of the American Aberdeen-Angus Breeders' Association, writes: "We are now recording more animals per annum than the parent society in Scotland, and our sales are also greatly in excess of those made in Scotland." This is worth thinking about.

GALLOWAYS.

Many judges have found Galloway and Angus so much alike that the one was mistaken for the other. This has been carried to such a point that claims were made at one time that there was no difference between them. In appearance their common color makes identification difficult, but as to a common origin, if such was the case, it was so far back in bovine mythology that its effect was long since lost owing to differences in climate, environment, management, and standards of breeding.

We find the Galloway ranking with the Angus in size. In form the beef rectangle presents itself—broad, deep, and symmetrical. Possessed of ample bone, very hardy and an ideal hustler, the Galloway has endeared himself to cattle raisers in rigorous climates and high altitudes. His coat is specially characteristic, with a thick hide and a mossy coat of long, wavy hair, in fact, a sort of double coat, a close mat of short hair being found under the long one. Black, tinged with brown, is the prevailing color, but a tendency to reversion to the old stock of mixed colors may be present.

Like the Angus, Galloway bulls are good dehorners, and strongly prepotent, from 75 per cent to 90 per cent of their calves from horned cows being without horns. On the range and in the feed lot Galloways are of great value, and their hides are made into robes of great beauty.

The principal defects that Galloway breeders are endeavoring to breed out at present are the roughness of the shoulder points and tail head, and an occasional lowering of the top line. Among the more outstanding merits of this breed are their great hardiness, prepotency, excellence for the block, and fine hides.

DEVONS.

Small in numbers and below medium in size, the "little Devon" is a bovine exemplification of the familiar bit of homely philosophy that articles of much value are often found in small packages. However, our Devon is not popular, according to general acceptance of the word, and, since the decadence of the ox as a motive power, has lost yet another claim on public affections. The reason for this is probably his size. Mature bulls do not often weigh over 2,000 pounds; but for high quality, fattening powers, compactness of flesh, and perfection of form, splendid handling, and beauty when finished, the Devon is a model.

In color, the Devon is a red varying from light to dark, with orange rings around the eyes, and tip of tail white, and white occasionally on the udder. It has a small, neat head, with broad forehead, and fine horn, curving forward, upward, and outward in cows and heavier and straighter in bulls. The body is set on short legs with fine bone. With all their value for beef production, the milking qualities of the Devons have not been allowed to deteriorate, which gives them special importance with the small farmer in a system of diversified agriculture. They are very hardy and great rustlers, and their value in regions of scanty pastures follows as a matter of course.

It is generally believed that the Devon has an ancient lineage that might startle some of his more fashionable brothers if it were disclosed, and, if the cattle that were brought over in the ship *Charity* really were unimproved Devons, it has the oldest record of any breed now on this continent. Though several importations were made prior to 1817, the one of that year was the first whose descendants were kept pure. From that time the growth of the breed has been slow. The Devon bulls are used to some extent in the range country, especially in Texas, giving much satisfaction. Their great value there would seem to warrant the extension of the breed on our lighter pastures.

SUSSEX.

This breed is not well known in the United States. It resembles the Devon closely, but is much larger, ranking nearly with the Shorthorn and Hereford. The color is much the same as the Devon, with brush of tail gray, and white markings on the underline and on inside of flanks.

The oxen of this breed have always been famous for strength and activity, and the steers make excellent beef. The record of the Sussex at the American fat-stock shows of ten and twelve years ago show them to possess qualities that well repay attention and care.

RED POLLS.

The Red Polled is one of the youngest of the breeds. It was not until the year 1846 that the union of the Norfolk and Suffolk breeders gave the breed its name, and from that time the real history of the breed may be said to date, though for many years before this the two branches had preserved their individuality in their respective districts. From the start this breed has been famous as one valuable alike for dairying and for beef production; and on the markets of England the Norfolk cattle take high rank, often selling for prices as good as the "Scotch" beef, which is considered the best the market affords.

They were first brought to America in 1873, and since have grown steadily. Here their beef-making qualities have been neglected, to some extent, but they have figured strongly as valuable animals for the small farmer. Neglect to show them in high condition has also tended to obscure their flesh-bearing powers in time past, but recent exhibitions have been highly creditable.

As their name indicates, cattle of this breed are without horns—no appearance of them being tolerated. In color they are a rich, deep red, with white allowed on the udder and underline, inside the flanks, and on the switch of the tail. The head is quite characteristic, of medium size, clean cut, with a sharp poll which carries a good tuft of hair. The neck is of medium length; body of good size and shape;

legs of medium length. Red Polls are very uniform, prepotent, and hardy, and have many earnest advocates, being good milkers as well as good feeders.

POLLED DURHAMS.

These cattle have quite reached a point where their claims as a breed are meeting recognition. Starting with a cross of Shorthorn bulls on native "muley" cows, and following by the use of hornless calves bred in registered Shorthorn herds, the polled feature has become fixed, and the bulls are very prepotent, the great majority of calves being without horns.

Aside from the absence of horns, the breed has most of the characteristics of the best Shorthorns. In fact, the model of the breeders is the typical Shorthorn. Many animals are eligible to registry in the American Shorthorn Herd Book as well as the American Polled Durham Herd Book, and in such cases are known as "double standard" Polled Durhams. Such animals have, of course, descended from purebred Shorthorns on both sides. Animals tracing to the old "muley" stock are not eligible to Shorthorn registry, and are called "single standard."

The breeders of these cattle are reasonable in their claims, and have shown commendable energy in developing their business, several animals having been shipped to South America.

BROWN SWISS.

These cattle are readily recognized by their color. This is a solid brown, varying in shade from light to dark, and varying in intensity on different parts of the body—the head, neck, legs, and quarters being usually the darker portions. The characteristic markings are a light, mealy ring around the muzzle, a light stripe across the lips and up the sides of the nostrils, a light-colored tuft of hair on the poll, and a light stripe down the back from shoulder to tail head, with black nose, tongue, switch, and hoofs.[a]

Brown Swiss cattle are of medium size, rather coarse, but hardy and rugged. Although developed almost exclusively for the dairy, they fatten well and have considerable value as beef.

From their native canton, Schwytz, they have been disseminated widely throughout Europe. They were first brought to the United States in 1869.

POLLED HEREFORDS.

In 1898 Gen. W. W. Guthrie, of Atchison, Kans., showed a group of cattle at the Trans-Mississippi Exposition at Omaha that attracted much attention. These cattle had Hereford characteristics, minus the

---

[a] Condensed from Report of Bureau of Animal Industry for 1898.

horns, and General Guthrie called them "Polled Kansans." They had been produced by crossing Hereford bulls on "muley" cows, inbreeding being avoided by resort again to Hereford stock. The bull shown was said to get a good percentage of hornless calves. Since then the interest has grown, the name Polled Hereford has been adopted, and recently the American Polled Hereford Cattle Club was organized, with headquarters in Des Moines, Iowa; Warren Gammon, secretary.

The polled variation is not so common among Herefords as among Shorthorns, and this imposes a considerable disadvantage on one who attempts to fix it; but there are occasional registered Herefords that have never developed horns. An effort is now being made to collect these animals into one herd and thereby form the nucleus for more extended operations. Such work is commendable and will be watched with interest.

PEDIGREES.

"A pedigree is the written record of an animal's ancestry." Every "scrub" has, in a certain sense, a pedigree, as well as an animal with a "title clear" on the page of a herd book, but with this difference, that the pedigree of the one has been without control or record for generations, while that of the other has been carefully kept in writing, and therefore fills the technical definition. The registered pedigree shows descent through animals of a certain standard and recognized merit, selected with care and skill, while only the course of nature has influenced the breeding of the "native" stock.

Two forms of pedigree are in use in this country for cataloguing sales—the condensed form and the extended or tabulated form.

THE CONDENSED FORM.

The condensed form is by far the more universally used. It is used in some form by the members of practically every live-stock association. A copy from a recent Shorthorn sale catalogue is given:

Buttercup.

[Roan; calved April 15, 1898; bred by E. B. Mitchel & Sons. Vol. 45.]

DAMS.	SIRES.	BRED BY—
	Baron Cruickshank 3d 117968	C. B. Dustin
12th Phyllis of Cedar	Goldstaff 92587	Robert Miller
9th Phyllis of Cedar	Fennel Duke 46070	G. Allen & Son
6th Phyllis of Cedar	6th Duke of Sharon 29364	A. Renick
1st Phyllis of Geneva	Geneva's Airdrie 19839	A. Renick
Twin Phyllis	Master Butterfly 14878	Saml. Thorne
Princeton Bell	Princeton 4285	R. A. Alexander
Daphne	Pearl 2012	S. Vanmeter
One Horn	Wellington (42586)	A. Renick
Limping Kate	Imp. Gratz (11542)	Dolley & Gratz
Susan Turley	Cossack (3503)	Mr. Clay
Katherine Turley	Bulmer (1760)	Earl of Carlisle
Imp. Young Phyllis	Fairfax (1023)	Mr. Whitaker

DAMS.	SIRES.	BRED BY—
Phyllis	Harpham (1098)	W. Cattle
Beauty	Percy (1312)	John Cattle
Delicacy	Ketton (346)	C. Colling

This is a cow of wonderful substance and great depth of flesh. She was shown successfully in the herd of E. B. Mitchel & Sons last year, standing first at Milwaukee and tied her stable companion, Rose Princess, for first place at Kansas City, the referee finally placing her second to the red heifer only.

BARON CRUICKSHANK 3d, probably well enough known to need little comment from us. Suffice it to say that as a breeder of early maturing, thick-fleshed calves, he has proven himself all that could be desired.

GOLDSTAFF, a bull of true Cruickshank type, combining a fine disposition with wonderful individual excellence, was sired by Imp. Goldstick 86748; out of Beauty's Pride 2d, by Orange Lad 46676; second dam, Beauty's Pride, by Imp. Baron Surmise 47432.

FENNEL DUKE, a Bates bull used at the head of Cedar Farm, got by Imp. Kirklevington Duke (41768), out of Imp. Princess 2d, by 6th Duke of Wetherby (33756).

6th DUKE OF SHARON, a pure Rose of Sharon, by Airdrie 3d (32919); out of Rosebud 13th, by the 4th Duke of Geneva (30958); g. dam Rosebud 7th, by 13th Duke of Airdrie (36459).

GENEVA'S AIRDRIE, by Airdrie 3d (32919); dam Nora Belle, by 2d Duke of Geneva 5562, and out of the Abram Renick Imp. Rose of Sharon.

Bred February 24, 1901, to Baron Montrath 127799.

The animal's name, of course, heads the pedigree; then follow the registration, description, date of calving, and name of breeder. The name of the sire of the animal is the name at the top of the middle column, while its dam's name is at the top of the left-hand column, and the name opposite this one is that of the sire of the dam—the maternal grandsire of the animal whose pedigree is under consideration. To illustrate, we have here the Shorthorn cow Buttercup (vol. 45, American Herd Book), got by Baron Cruickshank 3d 117968, out of 12th Phyllis of Cedar. The sire of 12th Phyllis of Cedar is Goldstaff 92587, and her dam 9th Phyllis of Cedar, whose sire is Fennel Duke 46070, and dam 6th Phyllis of Cedar. Thus the ancestry is traced through the dam, the dam giving the family name.

The names at the right are those of the breeders of the sires. In many cases the breeders of the dams are given at the left. This is of great advantage, for the quality of work done and reliability varies with different breeders, just as with producers in any other lines of business. Some advertisers omit the names of breeders entirely. After the pedigree, a number of notes are usually appended, showing the record of the animal in the show ring, and the breeding and record of the more prominent sires in the pedigree. This adds very much to its clearness, and tends to compensate the disadvantages of this form. If the pedigree is that of a cow, the date of breeding and name of bull are given, thus supplying data whereby the offspring may be registered.

The chief advantages of this form of pedigree are brevity and convenience. While it shows the line of family descent, it shows nothing of the breeding of the sire, and really very little concerning the dam. To estimate such a pedigree correctly, an amount of information is required that the majority of men do not possess. Showing the names of the breeders is of value, but if some are given, all should be, and this is obviously impossible. Such a form is not a pedigree in the true sense. It is incomplete and misleading, giving an impression of great merit when, perhaps, none exists. The pedigree above traces the maternal line of descent through fifteen generations. By a simple calculation, we find that the blood of 32,408 animals enters into the makeup of the progeny of so long a line of descent. Yet the condensed form shows but 30, with one sire, Ketton (346), from the sixteenth generation. In most cases, an infusion of foundation blood, by means of inbreeding, will have increased the percentage of such blood, yet the influence thus exercised must not be overestimated.

THE TABULATED FORM.

This style of pedigree has been coming into vogue within the past ten years, and is unquestionably the more rational. In ordinary sale catalogues, pedigrees are usually traced in the tabulated form through five generations, anything further being inconvenient and cumbersome. To show the difference between the two, the first five generations of Buttercup are tabulated and shown on page 25.

This shows at a glance the immediate lines of descent and reveals the presence of some of the greatest sires of Shorthorn history. Field Marshal and Roan Gauntlet might never be expected from the orthodox form, yet their value as sires in Buttercup's pedigree is much greater than that of Young Phyllis as a dam, and has had infinitely more to do with making this cow the excellent animal that she is. The chances of atavism—"throwing back"—to animals beyond the fifth generation are extremely slight with such strong top crosses, and the foundation must be made up of especially powerful and impressive sires and dams to be of direct value or detriment.

The only objection to the tabulated form is that it does not run the dam's descent to the bottom, and lacks the convenience of the "stepladder" form. But as this line and many more can be easily indicated in a tabulated pedigree by footnotes, that may be supplemented by a knowledge of the history of the breed, the increased value of this form over the other to trace foundation animals is obvious. As to the second point, one may well question whether accuracy should be sacrificed to convenience.

WHAT CONSTITUTES A GOOD PEDIGREE.

ELIGIBILITY.

Examining a pedigree, the first thing to be ascertained is whether the animal is registered or, if not, whether it is eligible. If not already registered, both sire and dam must have been, thus entitling the animal to registry. Having satisfied ourselves on this point, we next look for the name of the breeder.

Buttercup (vol. 45, A. S.-H. H. B.); roan; calved April 15, 1898; bred by E. B Mitchel & Sons.
- Baron Cruickshank 3d 117968.
  - Baron Cruickshank 106297.
    - Collingwood 106881
      - Pro-Consul 94510
        - Feudal Chief 92299.
        - Violet Bloom, by Royal Violet (40649).
      - Cardamine
        - Cumberland 50626.
        - Cowslip, by Barmpton (37763).
    - Maria 10th (vol. 35, p. 345, E.).
      - Field Marshal 64894
        - Roan Gauntlet (35284).
        - Azalea, by Cæsar Augustus (25704).
      - Maria 9th
        - Cleveland 106862.
        - Maria 6th, by Lord Paramount 108831.
  - Imp. Victoria 79th (vol. 35, p. 383).
    - Dunblane 65995
      - Roan Gauntlet 45276 ..... See below.
      - Duchess of Gloster 21st (vol. 26, p. 392, E.).
        - Barmpton Prince 45247.
        - Duchess of Gloster 13th, by Grand Duke of Gloster 19900.
    - Victoria 74th (vol. 32, p. 313, E.).
      - Roan Gauntlet 45276
        - Royal Duke of Gloster (29864).
        - Princess Royal, by Champion of England (17526).
      - Victoria 57th (vol. 26, p. 393, E.).
        - Barmpton Prince 45247.
        - Victoria 41st, by Lord Privy Seal 58649.
- 12th Phyllis of Cedar (vol 37, p. 658).
  - Goldstaff 92587
    - Goldstick 86748
      - Chancellor 68693
        - Barmpton (37768).
        - Crocus, by Pride of the Isles (35072).
      - Geranium (vol. 26, p. 392, E.).
        - Pride of the Isles 45274.
        - Garland, by Scotland's Pride (25100).
    - Beauty's Pride 2d (vol. 30, p. 760).
      - Orange Lad 46679
        - Royal Barmpton 31461 (32996).
        - Lovely, by Stanley 21127.
      - Beauty's Pride (vol. 30, p. 760).
        - Baron Surmise 47432.
        - Queen of Beauty 2d, by Stanley 21127.
  - 9th Phyllis of Cedar (vol. 34, p. 705).
    - Fennel Duke 46070
      - Kirklevington Duke 46385.
        - Duke of Wetherby 6th (38756).
        - Lady Kirklevington 2d, by Earl of Oxford (21655).
      - Princess 2d
        - Duke of Wetherby 6th (38756).
        - Princess, by 2d Duke of Wetherby (21618).
    - 6th Phyllis of Cedar (vol. 27, p. 664).
      - 6th Duke of Sharon 29364
        - Airdrie 3d 13320.
        - Rosebud 13th, by 4th Duke of Geneva (30958).
      - 1st Phyllis of Geneva (vol. 21).
        - Geneva's Airdrie 19839.
        - Twin Phyllis, tracing to imp. *Young Phyllis*, by Fairfax (1023).

The letter "E." after a cow's record refers to the English Herd Book.

THE BREEDER.

It is a matter of the greatest importance that the breeder be a man of recognized ability and unquestioned integrity. There is so much dependent on a man's method and management that the first point is of vital importance, and the opportunities for errors or downright falsehood are so great that the least question as to honesty should be carefully heeded. The life of a breed depends upon the honesty of the men who make out the pedigrees, and there should be severe punishment for a man who deliberately falsifies a pedigree or certificate of breeding. It is due cattle breeders, however, to say that their transactions have been remarkably free from any tampering with pedigrees.

MERIT.

After registry is looked to and we are sure that the animal comes from the hands of a breeder of recognized standing, we come to the pedigree itself. This is where knowledge of the breed is required, and discernment needed to know not only the character of the animals as individuals, but the results that will probably follow their mating. As "like begets like or the likeness of ancestors," valuable animals in a pedigree give a reasonably certain promise of a good animal as the offspring.

The first essential is to note well the qualities of the sire and dam, their record in show ring and breeding herd, and the records of their offspring. Note, too, whether they are of such lines of breeding that they will mate well, it being supposed that the animal has not yet been seen and examined. Trace the dam as far as possible, noting the breeding and showing qualities of her ancestors, for breeding qualities are hereditary as well as others. In examining a show record the particular shows in which winnings were made and the years are important, for a beast may win at one show and lose at another in the same season, and the same show will often vary in strength of competition from year to year. Take the line of the sire in the same manner, looking to it that strong animals make it up, and that it rests on a substantial foundation, for the mention of remoteness of ancestors spoken of above must not be taken as an argument that the base animals have no value. The thing to be desired above all others is a pedigree resting on a foundation made up of animals of great merit with a superstructure in harmony with it. The length is valuable when the pedigree traces through animals of merit in the hands of able breeders, each additional year fixing the characteristics of the breed and adding to prepotency. The better the foundation the better the opportunities of the breed.

The essentials of a good pedigree are, then, (1) eligibility to registry, (2) reliability and skill of breeders and their predecessors, and (3) individual merit of animals composing it.

FAMILIES.

Families in live stock are always more or less puzzling. Though the system of naming them is simple enough, the application sometimes appears to be incongruous. There is no material difference, however, between our system of naming ourselves and the way breeders name their animals, except that in the latter case the family name is not always appended. A man may be called "Smith" when he really has more blood in his veins of the Jones family, but as is the father's name, so is the son's. With bovines, with a few exceptions, the family name is that of a cow who has had a great record or gives promise of becoming a great one or whose offspring are of uniform and outstanding excellence. All her descendants are spoken of as belonging to that particular family. For example, we have the cows Young Mary, Young Phyllis, Duchess, Secret, Duchess of Gloster, Lavender, Princess Royal, etc., among the Shorthorns, that have been arbitrarily taken as the founders of families. In the United States Shorthorn families usually trace to imported cows. Among the many Angus cows which breeders have chosen to found families may be mentioned Pride of Aberdeen, Queen Mother, Kinnochtry Princess, Blackbird of Corskie, Coquette, Erica, etc. The Hereford breeders are not sticklers for family trees, but have a general preference for certain lines of blood that are traced to the sires by no special route, i. e., the "family line," if we may call it that, may run through the sires as well as the dams. A Hereford bull, then, may found a family, though the lines are not drawn hard and fast, and his descendants may spurn their lineage. Among the bulls of this breed whose blood is popular at present are Anxiety and his sons, The Grove 3rd, Lord Wilton, Corrector, Beau Real, and Hesiod.

THE VALUE OF A GOOD FAMILY.

There are unquestionably good reasons for breeding animals within blood lines that have proved themselves to have exceptional merit in show ring or feed yard. There are, indeed, worthless families of cattle, but, since custom does not require them to be held up to the eye by means of a surname, they may escape notice. In every breed of cattle there are animals that are undesirable in all respects. Of all scrubs, the pedigreed scrub is by far the worst, for someone is sure to breed from it and pass the offspring on to his neighbor with the assurance, "It's all right; it has a pedigree." By keeping to approved lines of blood and requiring individual excellence in each animal such conditions may be avoided, though it must be understood that the breeding of a bovine is only one factor in his destiny. Feed, care, and environment are others no less important.

The great danger of an adherence to family lines is that the pursuit of a fashionable pedigree may become the sole aim of the breeder's work. This has happened, and when such methods prevail the breed will be in very serious danger.

To breed animals entirely within the limits of one family, as Shorthorn men once tried to do in their pursuit after "pure" Duchesses, and as a few attempt now in a desire to possess "straight" Cruickshanks, is to resort to very violent and dangerous inbreeding; and as this is an "edged tool" that can be used only by a master, and as the master minds usually steer clear from "straight" and "pure" hallucinations, it is evident that "family" may be a danger signal instead of a beacon. On the other hand, calling an animal by a family name long after all family resemblance is lost through the infusion of outside blood and nothing but the name is left is a delusion that is much more inexcusable, if not as dangerous as the pursuit of the "straight" pedigree phantom.

A good family is desirable, but the essentials in securing it are, first, to see that the pedigree does not mislead and place too great importance on a distant ancestor; second, that high individual excellence goes hand in hand with a good family descent.

REGISTRATION.

To provide a means whereby pedigrees may be recorded, the history of a breed accurately and officially chronicled, and its integrity and identity as a breed preserved, herdbooks are published. They vary more or less in minor details, but the fundamental principles of all are the same. In the early days breeders were not organized as at present, and the work of compiling a herdbook and recording pedigrees was usually the effort of some faithful and gifted lover of the breed who sacrificed time, labor, and often means to such work. Now, in nearly every instance the management of the herdbook is the chief business of an association of breeders, the editing of such being placed in the hands of the secretary. Details relating to routine business, shows, and the general advancement of the breed, while of great importance, are secondary to the one great essential.

More or less detail and exacting requirements in the form of pedigree and method of recording bring about an occasional cry of "red tape." But rules must be made and rigidly enforced if the records are to be of any value.

CERTIFICATES OF REGISTRY.

The first step taken to record an animal is to fill out the "entry blank" furnished by the association, stating sex, name of animal, color, date of calving, name of breeder, names, registry numbers, and breeders of sire and dam, and often of grandsires and granddams, thus making the statement more complete and giving a convenient means of ascertaining its veracity at headquarters. This is forwarded with the proper fee to the secretary's office. Here the breeding is verified, entered upon the records, a number assigned the animal bearing it, and a certificate of registry is sent the owner, stating that the animal, calved and bred as stated in the application, is recorded by the number given in the association's herdbook, except with the Red Poll Association. Numbers are assigned consecutively as applications for entry come in. This number is then the means by which an animal's identity and pedigree may be established at any time, and is of vital importance, the name being only secondary, as there are many animals recorded under the same name.[a]

MISCELLANEOUS DOCUMENTS.

Certified pedigrees.—If desired, a certified pedigree may be issued with the certificate of registry. This is usually tabulated, certified as correct, and bears the signatures of the president and secretary of the association. It stands as an official record and is valued accordingly.

Transfers.—Applications for transfer are made out on sale of animals, stating sex, name, and number of animal, and names of seller and purchaser, and are used only with animals already recorded. They are usually sent by the seller to headquarters for record and a certificate of transfer is issued by the association.

Breeding certificates.—Breeding certificates and statements certifying to lease of sire are also used on occasion and are of great value to prevent entanglements in the ramifications of breeding operations.

Fees.—Nominal fees are charged for the various services just under discussion. Entry fees vary with the age of the animal, a penalty in the way of an increase of the fee attaching to neglect to register before certain ages. Fees for transfer are paid by the person sending them for record.

While the same general forms are used by all associations for animals of both sexes, we meet with numerous modifications and variations. For example, if a cow is registered, space is usually provided on the certificate of entry for date of breeding and name of bull. This accompanies every certificate given the purchaser on sale of a cow, and serves the same purpose as a statement of breeding in a sale catalogue. Some associations supply a different form of entry blank for animals imported from Great Britain, and in others the record of the females in the breeding herds is sent in annually. In some cases a cow may be rerecorded, this being done to show her produce.

---

[a] Volume XXII of the American Hereford Herd Book records 22 animals under the name of Daisy and 27 as Beauty, and the popularity of these names has recorded 30 "Beautys" and 32 "Daisys" in the 43d volume of the Shorthorn Herd Book, with variations innumerable.

THE HERDBOOK.

A perusal of pedigree to any extent leads one inevitably to the herdbook. These records are usually confined to mere presentations of the facts stated in the owner's application for entry, with no attempt to give more of an animal's pedigree than the names of his sire and dam. In some of the older volumes one will find that records are entered at greater length.[a]

To trace out a pedigree by means of the herdbook, all that is necessary is to find the record of the animal whose pedigree is wanted, thus obtaining the names and numbers of the sires and dams, looking up the breeding of these in the same manner and carrying the process out for each line of descent. By writing down the ancestors as they occur, we have a tabulated pedigree.

A representation of the forms employed by the association will serve to make this explanation clearer and may be of interest.

SHORTHORN.

All but the Shorthorn herdbook register animals by number, regardless of sex. Shorthorn bulls are registered by number, but the page and volume of the herdbook on which the breeding of a cow appears serves as her registry, and her lineage is always traced by means of the sire; e. g., we have Princess Alice (vol. 35, p. 628) by Field Marshal 64894; Mary Abbotsburn 7th (vol. 39, p. 612) by Young Abbotsburn 110679, etc. In describing a Shorthorn cow, therefore, the name and number of her sire are always essential to accuracy and completeness, for two cows of the same name are sometimes recorded on the same page. The principal advantage in recording a cow in this manner is that it provides a check whereby greater accuracy is insured.

The following is the form in which bulls are recorded:

137459   BARON BROWNDALE FOURTH.

Dark roan, calved July 8, 1897, bred by H. F. Brown, Minneapolis, Minn., got by Victor of Browndale 117621, out of Constance of Cloughdale 2d (vol. 39, p. 369) by Bloom's Duke of South Fork 90689—tracing to imp. *Constance* by Bridegroom (11203).

The number of this bull is 137459, his sire Victor of Browndale 117621, and his dam Constance of Cloughdale 2d, who is recorded on page 369 of volume 39 of the American Shorthorn Herd Book, and whose sire is Bloom's Duke of South Fork 90689. Baron Browndale Fourth traces through his dam to the imported cow Constance whose sire is Bridegroom, number 11203 in the English Herd Book.

---

[a] In addition to serving as the official repository of pedigrees, the herdbook is the official organ of the association, through which proceedings of meetings, news of the breed, and other matters of interest are presented to the public.

AMERICAN BREEDS OF BEEF CATTLE.

The following is the recorded form used for cows:

RUSSELLA—Red, calved March 20, 1894, bred by Ezra Swain, Noblesville, Ind., got by Czar 107007 out of Dutchess Nonpareil (vol. 37, p. 891) by Lord Nonpareil 63437—tracing to imp. *12th Duchess of Gloster* by Champion of England (17526).

1896, Dec. 14, red b. c., by Secret Victor 121092, Virginia C. Meredith.
1898, Oct. 14, roan c. c., Ruberta (vol. 45, p.) by St. Valentine 121014, J. G. Robbins & Sons.

This cow is recorded in vol. 44, page 854. The form is exactly the same as for bulls, with the omission of the number, and the difference in the position of the cow's name.

A produce table, showing the offspring of the cow, may be appended to her own record, as in this case. This shows the date the calf was dropped, its sex, sire, and breeder. Open numbers refer to the American Herd Book; numbers in marks of parenthesis ( ) to the English Herd Book. Cows are entered in the herdbook under the names of their breeders, which are arranged alphabetically. Names of imported cows are printed in italics. To be eligible to registry in the American Herd Book, all animals must have registered parents, and trace to an imported parent or be registered in the English Herd Book, imported animals tracing to cows registered in the first twenty volumes of that record.

HEREFORD.

As already indicated, all Herefords are numbered consecutively without regard to sex. Cows, therefore, are numbered, and their records are found in the herdbook by this means.

The following is a copy from a page of Vol. XXII, American Hereford Herd Book:

No.	Sex.	Name.	Breeder.	Dropped.
104321	Cow	Penelope	Thos. Clark	October 31, 1899.
104322	...do...	Polly Peachum	...do...	January 10, 1900.
104323	...do...	Bright Lass	J. A. Creed	November 6, 1899.
104324	...do...	Miss Kate Adams	E. W. Creed	October 15, 1899.
104325	Bull	Prince Silo	Frank Petz	November 22, 1899.
104326	...do...	Rob Roy	Geo. E. McEathron	February 24, 1899.
104327	Cow	Helen Mar 2d	J. E. Foss	March 8, 1900.
104328	Bull	Wonderful	...do...	January 9, 1900.
104329	Cow	Daisy	Wm. Bowers	March 22, 1899.
104330	...do...	Hannah E	Estate of A. Gregory	September 18, 1899.

No.	Sex.	Name.	Owner.	Sire.	Dam.
104321	Cow	Penelope	Thos. Clark	Le Roy 70778	Peach 40689.
104322	...do...	Polly Peachum	...do...	...do...	Patty Lee 55712.
104323	...do...	Bright Lass	J. A. Creed	Monitor 80954	Daisy Grove 80953.
104324	...do...	Miss Kate Adams	E. W. Creed	Jocko 66942	Kate Adams 50145.
104325	Bull	Prince Silo	Frank Petz	Dandy 72195	Ethel 42966.
104326	...do...	Rob Roy	W. S Billinghurst	Guardian 58555	Shy Lass 45426.
104327	Cow	Helen Mar 2d	J. E. Foss	Ezra 55999	Helen Mar 12412.
104328	Bull	Wonderful	...do...	Fortune, jr., 57153	Cassie 56384.
104329	Cow	Daisy	G. Hornaday & Co	Fortune jr. 57153	Mabel 70274.
104330	...do...	Hannah E	Wm. Humphrey	Duke of York 58674.	Miranda 2d 51799.

ABERDEEN-ANGUS.

The Angus Association records bulls and cows consecutively, publishing the pedigree in the catalogue form already shown (page 22). A sample is shown here:

26060.   QUEEN McHENRY 15th.   Cow.

Calved October 11, 1896. Black; some white on udder. Bred and owned by W. A. McHenry, Denison, Iowa.

		Heather Lad 4th 16747.	J. R. Harvey.
Gudgell & Simpson.	Jenny June 4775	Knight of St. Patrick 354.	R. C. Auld.
Peter Cran	Jenny Lind 4874	Marshal Var 613	John Hannay.
Peter Cran	Jemima 2d of Morlich 1607.	Champion of Findrack 439.	Captain Fraser.
Peter Cran	Jemima of Morlich 1470.	Patrick 440	Peter Cran.
Peter Cran	Fancy of Morlich 612	Balwyllo Eclipse 443	Trus. R. Scott.
Wm. McCombie	Beauty of Morlich 442	Angus 83	Hugh Watson.
Wm. McCombie	Windsor 107	Victor 132	Wm. McCombie.
Wm. Fullerton	Queen Mother 41	Panmure 69	Lord Panmure.
Wm. Fullerton	Queen of Ardovie 42	Captain 86	Mr. Sime.
	Black Meg 43		

GALLOWAY.

The Galloway form very closely resembles that used by the Shorthorn Association.

12957.   GRACE OF WAVERTREE—COW.

Calved March 21, 1890. Bred by Hugh Paul, of Dundee, Minn. Owned by G. H. Gurley, of Pipestone, Minn. Sired by Hero of Wavertree 3668, out of Imp. China 2d 2718 (4403) by Earl of Nithsdale (1035)—China (4402) by Lord of Nithsdale (616)—Meg.

This shows the cow Grace of Wavertree, recorded as 12957, her sire is Hero of Wavertree 3668, and her dam the imported cow China 2d 2718 (4403), by Earl of Nithsdale (1035). China 2d was recorded in the Scotch herdbook as (4403). Her dam was China (4402), by Lord of Nithsdale (616), and China's dam was Meg.

DEVON.

Devon bulls and cows are numbered separately. This presents some very commendable features, no possible danger of confusion presenting itself. The bull's record below is self-explanatory.

6183.   SPARTAN.

Calved February 1, 1894. Bred by Jos. Hilton & Sons, New Scotland, N. Y. By Lord Flemington 4773. Dam Sturdy Dame 7314.

RED POLL.

The Red Poll herdbook appears somewhat more complicated at first sight than the other herdbooks. Bulls and cows are entered separately, by numbers; the numbers of bulls being placed at the right margin, while those of the cows are at the left. In the pedigree itself, numbers of dams precede the name instead of following, as is usually the case. The foundation tribes are indicated by letters and families within these tribes by numbers. After the name of each female record is her tribe letter and family number. The tribe letter precedes the family number, except when pedigrees run to the old Norfolk and Suffolk stock instead of to the herd of some breeder of prominence, the numbers in such instances referring to the different herds of this old stock instead of individual animals. Thus, we will have A 1, C 4, M 2, but 1 NORF, 2 SUFF. "The earliest numbers are assigned to cows having the best recorded pedigree." The letters "s" and "d" in a pedigree stand for "sire" and "dam," respectively.

The form for bulls is as follows:

CRESCO LAD.   6923.

Calved October 2, 1899: breeder, S. A. Converse: s. Richland Boy 5th 5159 by Dobin 3462; d. 8672 Lida Falstaff 2nd by Red Skin 1278 by Davyson 7th 476; 2nd d. 5637 Lida Falstaff by Falstaff by Rufus 188; 3rd d. 1620 Lida—A 1 by Champion 271. Record to 7th d. 427 Primrose—A 1.

Tabulated, this pedigree would appear:

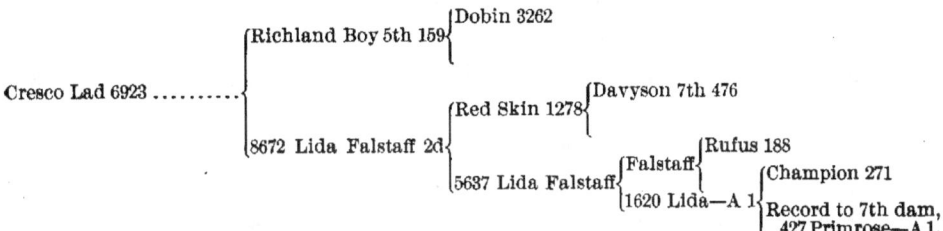

Cows are recorded under the following form:

14503.   BEELESS—1 NORF.

Calved August 20, 1899: breeder, V. F. Hills: s. Brutus 4275 by Lucre 3193; d. 6912 Bee by Delaware Chief 1781 by Duke of Dayton 663; 2nd d. 3248 Beatrice by Brutus Duo 463 by Brutus 269; 3rd d. 2577 Troublesome—1 NORF. Record to 4th d. Pond— 1 NORF.

POLLED DURHAM.

The Polled Durham herdbook creates much interest, especially when records of double-standard animals are read. The appended record is one of these:

165.   OTTAWA PHYLLIS DUKE.   116668.

Red, calved April 19, 1893, bred and owned by W. S. Miller, Elmore, Ohio, got by Ottawa Duke x 185 109292, out of Phyllis Lass 2nd, vol. 38, by 2d Duke of Barington, 50877—Phyllis Lass by Bates's Lad 28567—Gentle Annie 32nd by Airdrie 4th 11268— Gentle Annie 4th by Gay 8165—Gentle Annie 2nd by Tycoon 9236—Gentle Annie by Challenger 324—Princess by Prince Albert 2nd 857—Susan Turley by Cossack (3503)— Catharine Turley by Bulmer (1760)—imp. *Young Phyllis* by Fairfax (1023).

The number (165) at the upper lefthand corner, printed in bold-faced type, is that of the Polled Durham herdbook; that on the opposite side (116668) is the American Shorthorn number. This shows how closely the work of the Polled Durham approaches the Shorthorn plane. In addition to the hornless requirement, the presence of more and more Shorthorn blood is being required.

Pedigrees of bulls are inserted consecutively; those of cows under the names of their breeders, which are arranged alphabetically.

The cross (x) before a number in the Polled Durham herdbook indicates registry in that association, thus distinguishing it from the Shorthorn. The same rule is kept regarding English numbers as in all other associations, namely, placing them in parentheses.

BREEDERS' ORGANIZATIONS.

American Shorthorn Breeders' Association, John W. Groves, secretary, Springfield, Ill.

American Hereford Cattle Breeders' Association, C. R. Thomas, secretary, 225 West Twelfth street, Kansas City, Mo.

American Aberdeen-Angus Breeders' Association, Thomas McFarlane, secretary, Harvey, Ill.

American Galloway Breeders' Association, Frank B. Hearne, secretary, Independence, Mo.

American Devon Cattle Club, L. P. Sisson, secretary, Newark, Ohio.

American Sussex Association, Overton Lea, secretary, Nashville, Tenn.

The Red Polled Cattle Club of America (incorporated), J. McLain Smith, secretary, Dayton, Ohio.

American Polled Durham Breeders' Association, Fletcher S. Hines, secretary, Indianapolis, Ind.

American Polled Hereford Cattle Club, Warren Gammon, secretary, Des Moines, Iowa.

Brown Swiss Cattle Breeders' Association, N. S. Fish, secretary, Groton, Conn.

○

POLLED HEREFORD COW.

Photograph furnished by Gen. W. W. Guthrie.

Photograph furnished by Gen. W. W. Guthrie.

POLLED HEREFORD BULL.

POLLED DURHAM COW GOODNESS 15TH.

By courtesy of J. H. Miller.

Polled Durham Bull Golden Gauntlet.

Photograph furnished by J. H. Miller.

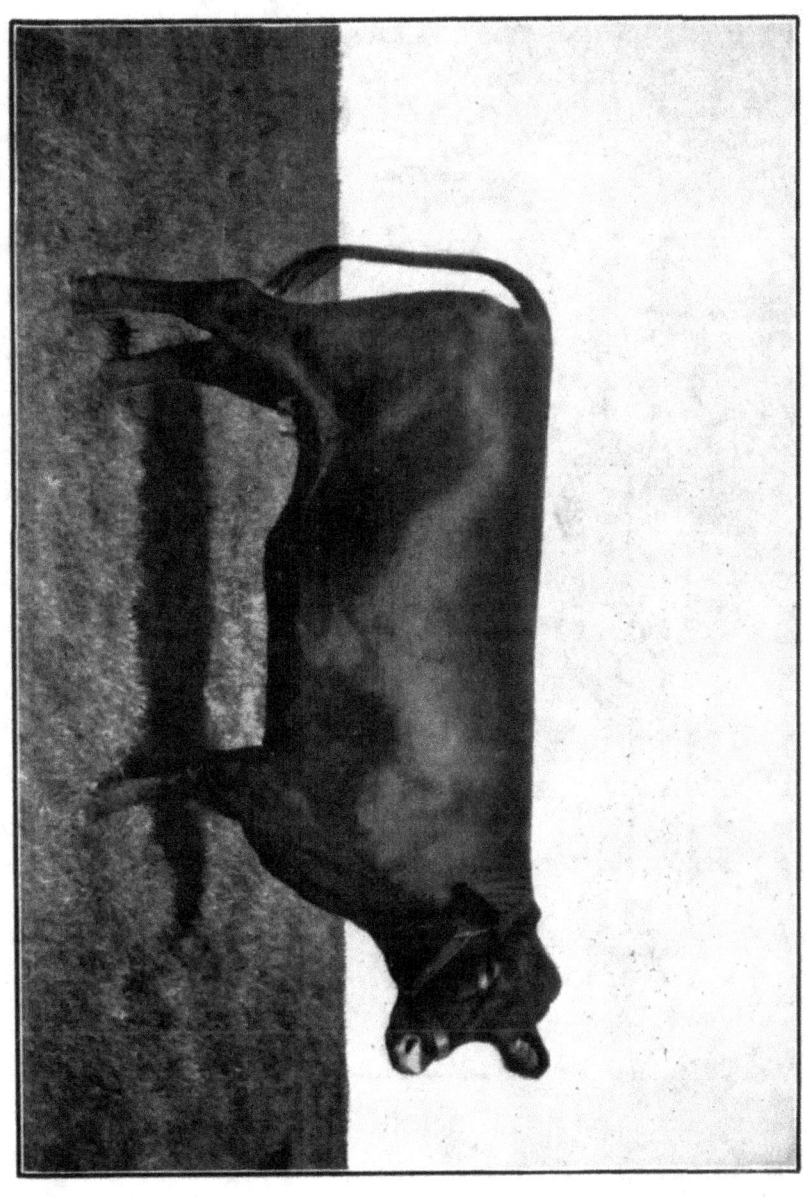

Red Polled Heifer Lida Falstaff 3d.

Photograph furnished by Wallace's Farmer.

Red Polled Bull Demon.

Photograph furnished by Andrew Bros.

Devon Heifer Tulip of Woodland 23d.

Devon Bull Tulip's Royal 1st.

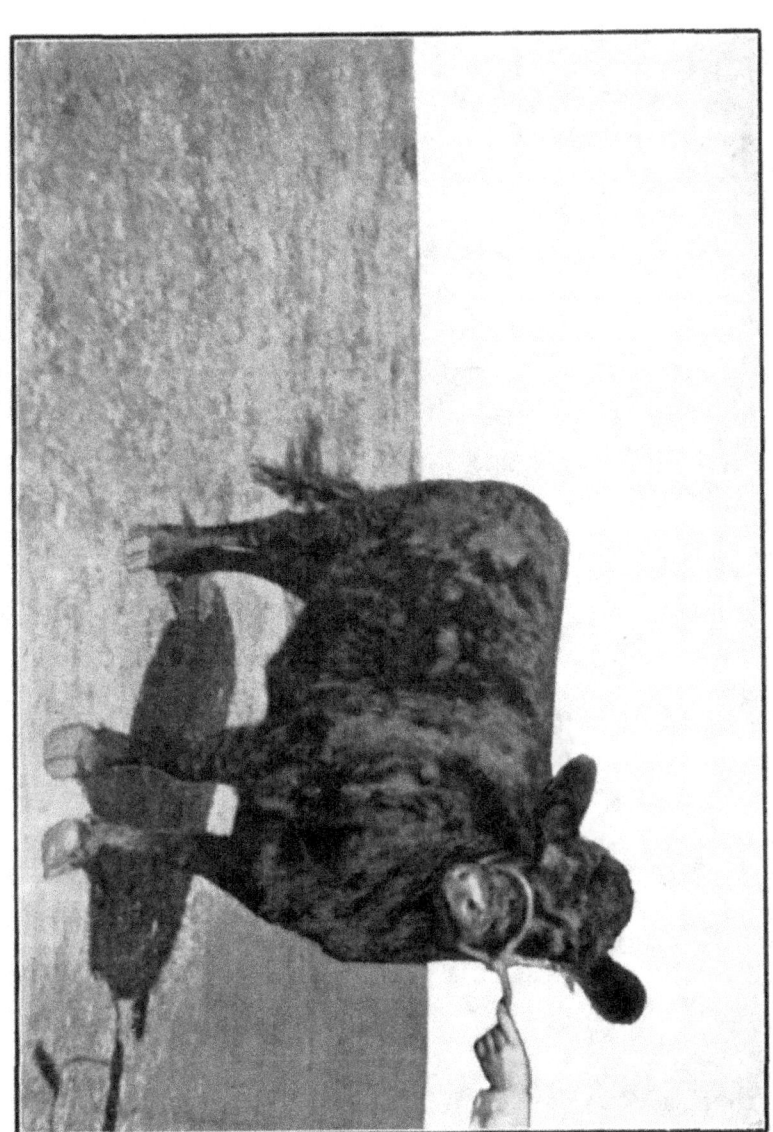

Photograph furnished by Wallace's Farmer.

GALLOWAY HEIFER LUTIE LAKE.

Galloway Heifer Gentle Annie A.

Photograph furnished by Wallace's Farmer.

Photographs furnished by A. C. Binnie.

FRONT AND REAR VIEWS OF ORIN OF LONGBRANCH.

Photograph furnished by Dr. W. H. B. Medd.

GALLOWAY BULL SPECULATOR OF DUNDEE.

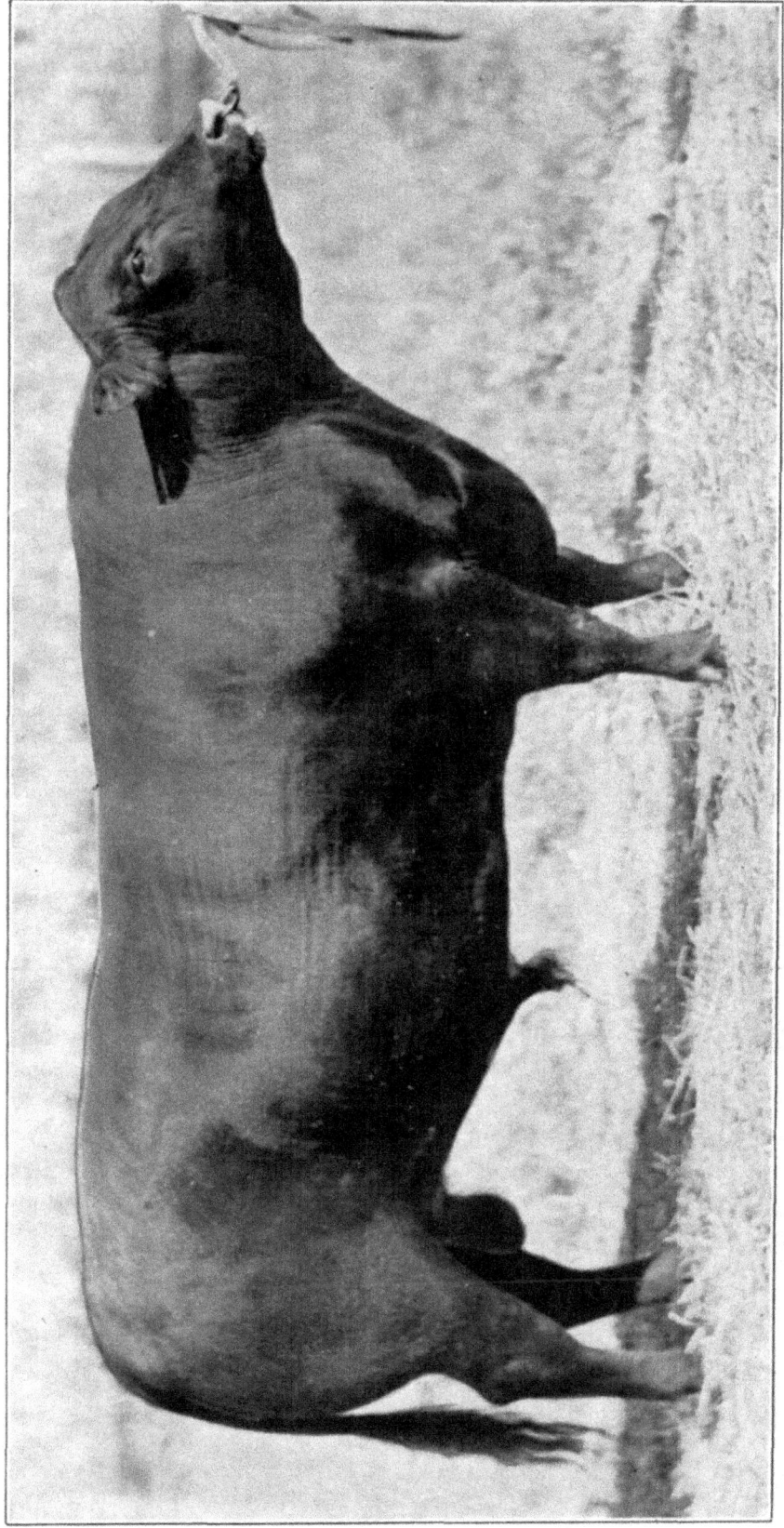

Photograph furnished by A. C. Binnie.

ABERDEEN-ANGUS BULL ORIN OF LONGBRANCH.

Hereford Cow Betty 2d.

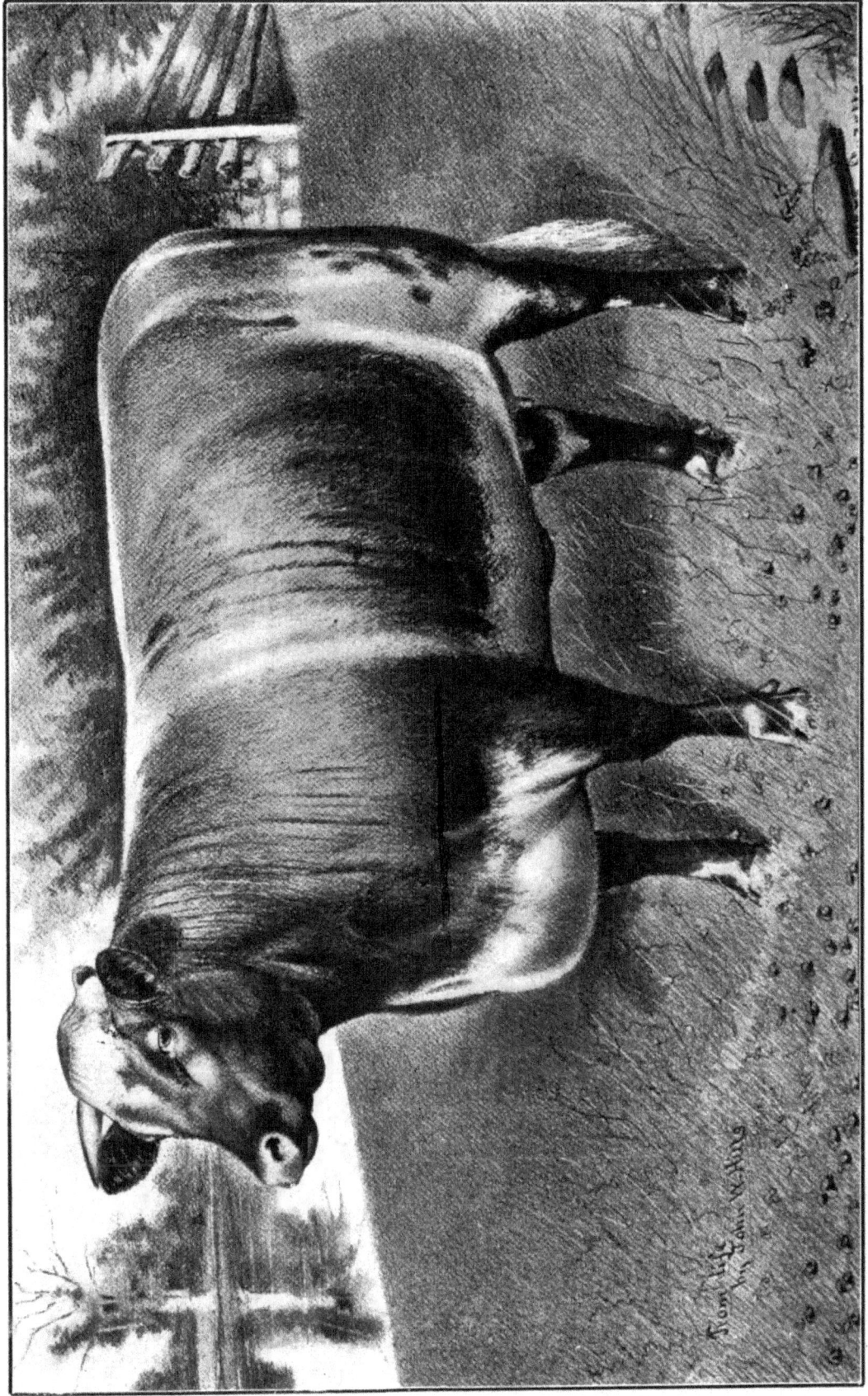

By courtesy of Aaron Barber.

SHORTHORN COW MARY ABBOTSBURN 7TH.
Bred by the late T. S. Moberly, Lexington, Ky.

Photograph furnished by T. F. B. Sotham.

REAR VIEW OF SIR BREDWELL.

By courtesy of T. J. Wornall.

REAR VIEW OF VISCOUNT OF ANOKA.

By courtesy of T. F. B. Sotham.

HEREFORD BULL SIR BREDWELL.

By courtesy of T. J. Wornall.

SHORTHORN BULL VISCOUNT OF ANOKA.

Photograph furnished by J. G. Robbins & Sons.

SHORTHORN HEIFER RUBERTA.

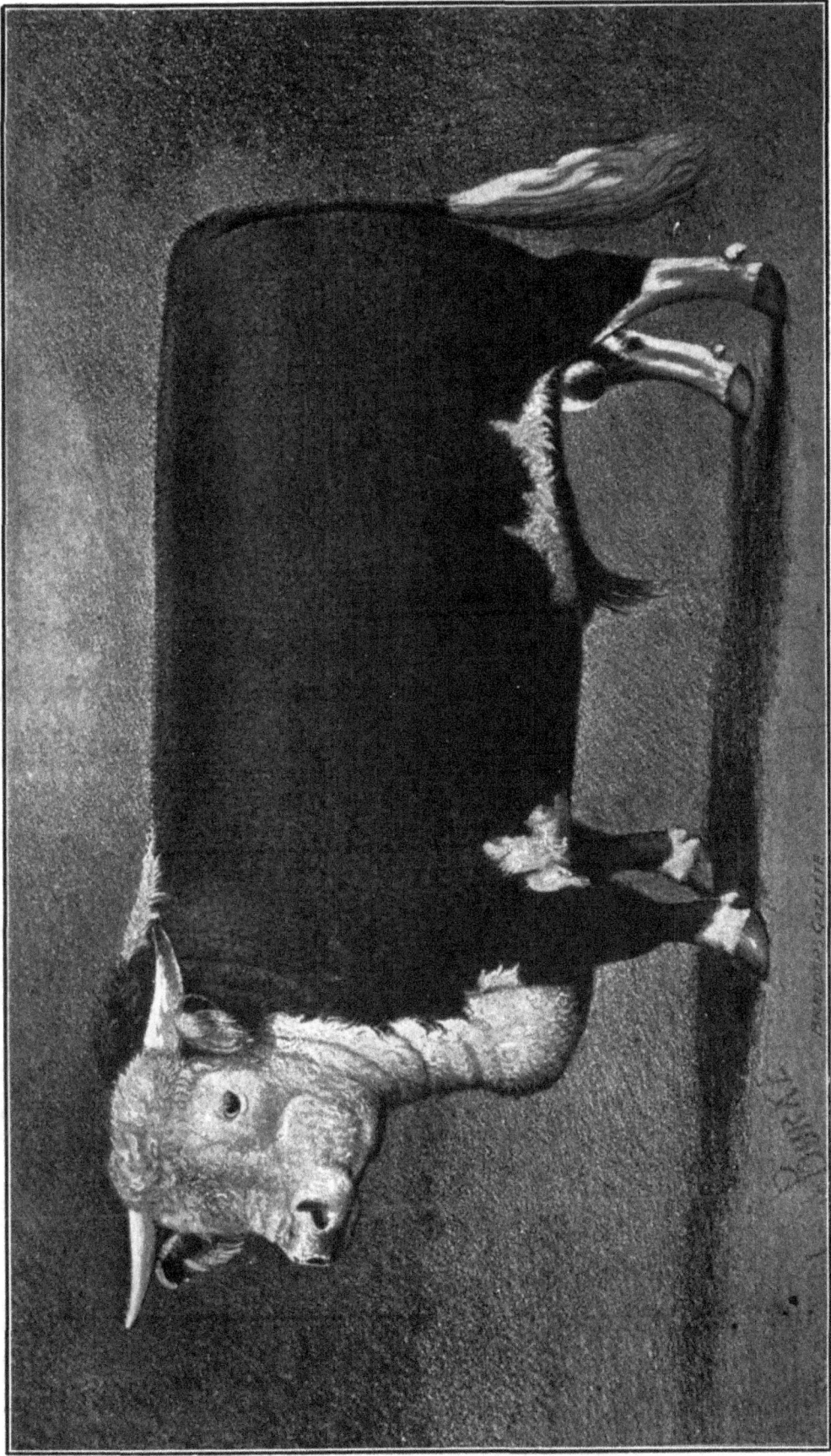

By courtesy of Clem Graves.

HEREFORD BULL DALE.

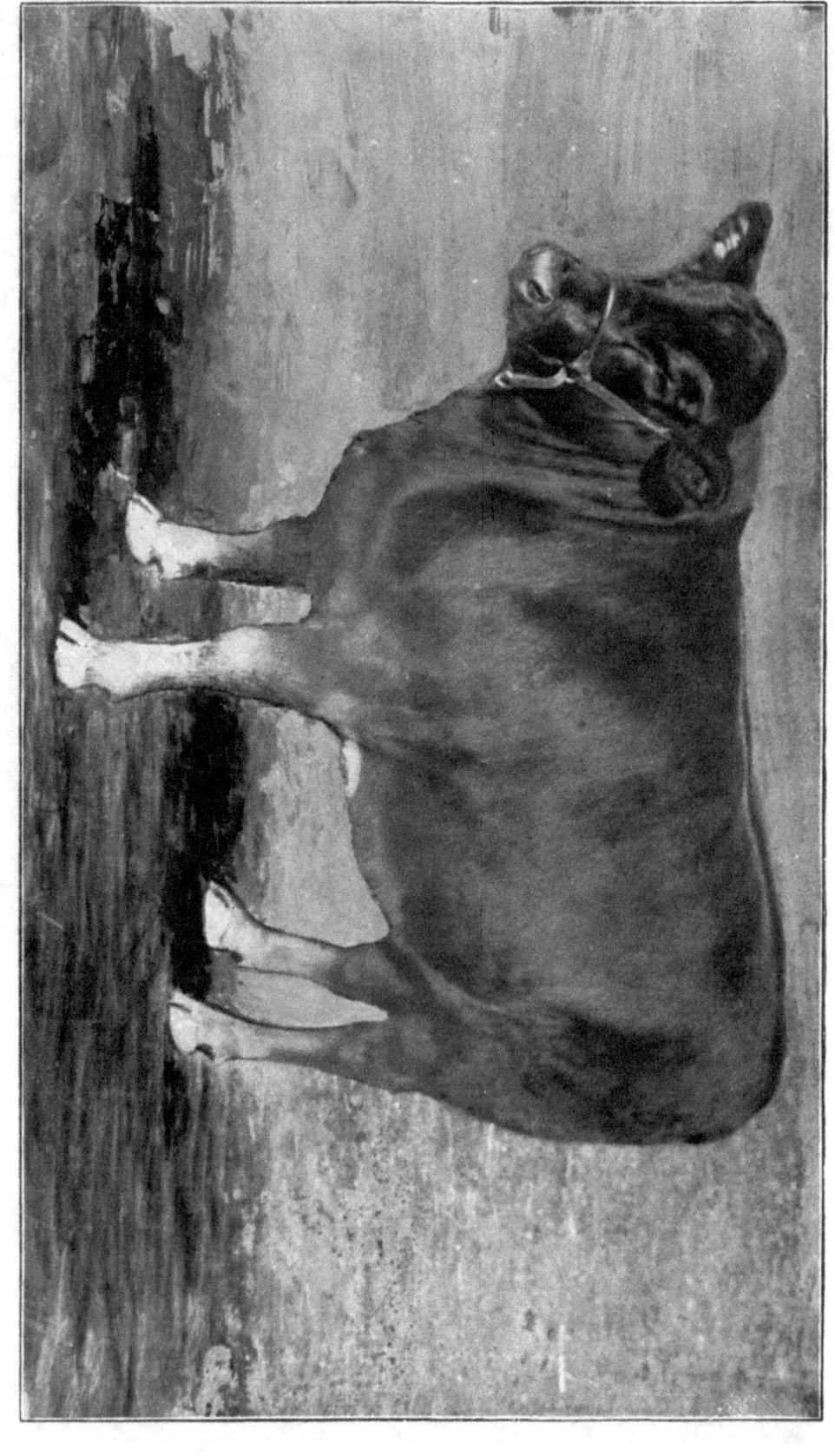

ABERDEEN-ANGUS ADVANCE.
Champion steer of International Live-stock Exposition, 1900. Bred, fed, and exhibited by Stanley R. Pierce, Creston, Ill.

By courtesy of Wm. Blackwood & Sons, Edinburgh.

THE BLACKWELL OX.

www.ingramcontent.com/pod-product-compliance
Lightning Source LLC
Chambersburg PA
CBHW060003230526
45472CB00008B/1934